ADDRESS

ON

PROGRESSIVE AGRICULTURE,

AND

INDUSTRIAL EDUCATION,

DELIVERED BEFORE THE

Mississippi Agricultural and Mechanical Fair Association, at Jackson, November 14th, 1872,

By EUG. W. HILGARD,

PROFESSOR OF AGRICULTURAL CHEMISTRY AT THE UNIVERSITY OF MISSISSIPPI, AND STATE GEOLOGIST.

JACKSON, MISS.:

PRINTED AT THE CLARION BOOK AND JOB OFFICE.

1873.

ADDRESS

ON

PROGRESSIVE AGRICULTURE,

AND

INDUSTRIAL EDUCATION,

DELIVERED BEFORE THE

Mississippi Agricultural and Mechanical Fair Association, at Jackson, November 14th, 1872,

By EUG. W. HILGARD,

PROFESSOR OF AGRICULTURAL CHEMISTRY AT THE UNIVERSITY OF MISSISSIPPI, AND STATE GEOLOGIST.

JACKSON, MISS.:

PRINTED AT THE CLARION BOOK AND JOB OFFICE.

1873.

ADDRESS.

I appear before you to-night with the view of presenting some considerations on Improvement in Agriculture, as connected with Industrial Education. If in so doing I confine myself to the subject of Agriculture proper, it is not because I underrate the necessity and importance of providing for the professional education of those pursuing the several branches of the Mechanic Arts; but simply because with us, at the present time, agriculture is the overshadowing interest, claiming the first consideration; and because the vastness of the whole subject would, were I to do it justice, impose too severe a tax upon your patience.

It might seem superflous to demonstrate the necessity of a serious change in our agricultural habits and practices. Yet there are too many who, though in general admitting this, fail to appreciate the pressing necessity, and the extent of the change required.

In an agricultural commonwealth, the fundamental requirement of continued prosperity is, beyond any possible cavil, that *the fertility of the soil must be maintained.* Whenever this condition fails, wholly or in part, of fulfilment, agriculture must, to a corresponding extent, cease to be the occupation of its inhabitants, especially if other countries compete with them in the same pursuit, under more favorable circumstances. The population must, in that case, turn to other pursuits, if the natural resources of their country permit them to do so while purchasing their supplies abroad—as happens, e. g., in mining regions. But when there is no such choice of pursuits, the result of the exhaustion of the soil is simply *depopulation;* the inhabitants seeking in emigration, or in conquest, the means of subsistence and comfort denied them by a sterile soil at home.

THE LESSONS OF HISTORY.

History, both ancient and contemporary, furnishes abundant examples of the working of these causes. The decline of empires and the decay of nations have so often gone hand in hand with the decline of the soil's productiveness, that the coincidence cannot escape the eye of any student of history. It was so in Greece and Rome; and neither Greece nor Italy have recovered from the depopulation resulting from the emigration of the most vigorous portion of their once teeming population, to regions possessing soils unexhausted, and offering a larger reward for their toil. What were once the most fertile portions of ancient Latium, are now wastes of grass and thistles, supporting but a sparse pastoral population; and the dreaded Pontine swamps were, at that time, the site of numerous thriving villages. The Georgics of Virgil, and the treatises of Columella, show that the same difficulties we are now beginning to experience, were seriously felt in their times; and the desolation of the once fertile Roman Campagna has its parallel in the gullied commons waving with broom-sedge, that surround most of our older country-towns.

Spain is another case in point; and as we are much in the habit of sneering at the "decline and fall" which that once potent empire has experienced, let us be sure to profit by the teachings of its history. Hispania was esteemed the most fertile province of the Roman empire; and in the reign of Abd Errahman III. (A. D. 961), Mohametan Spain alone counted some thirty millions of inhabitants.

Six centuries later, under the reign of Philip the Second, the Spanish writer Herrera says, in his treatise on agriculture:

"What may be the cause that now-a-days the deficiency of food makes itself felt in the whole land, and that now, in times of peace, a pound of meat costs as much as, not long ago, a whole mutton in the midst of war? Over-population cannot be the cause, for where a thousand Moors once found employment, there is now scarcely room for five hundred Christians. Neither can it be the importation of gold from India. *Is it perhaps the soil which lies dormant?* But the soil does not need any other rest than the winter's sleep; and there was no lack of winter rains to refresh it, and to provide it with force for the sprouting of seeds. What, then, is the cause that the soil will not nourish us any more?" And, like some of our modern believers in quack nostrums and panaceas, he answers: "The *mule* is the cause. In the 13th century it gained ground, since which time dates the desolation of Spain. It has not the strength to plow deep enough."

Doubtless, with deeper tillage, productiveness might have been longer maintained; even as with us, subsoiling is the first step towards the reclamation of worn soils. But the real fault lay in the idea, that "the winter's sleep" was sufficient to restore the soil's loss from cropping. It is perhaps to this mistake, that we owe the discovery of America, and of the sea-route to India; for, failing to make their living, much less their fortunes, at home, the enterprising part of the population sought them in the discovery and conquest of distant lands.

HOW WE REPEAT HISTORY.

And our own population is once more repeating history, under the influence of the same causes. Armed with better implements of tillage, it takes them but a short time to "tire" the soil first taken into cultivation; which is then turned out, while the fence is transferred to another tract, newly cleared. This in its turn is exhausted by continuous cropping, year after year, with the same cotton and corn. By this time, probably, our backwoodsman finds the neighbors getting too close to him for comfort, and his land and "improvements" are for sale at whatever price he can get for them; and the next winter finds him on his way to Texas or the territories—where, in time, he will repeat the same cycle of operations.

It is to the roving propensities of these hardy pioneers, that we owe the rapid development of the Far West, to whose conquest they are steadily advancing. But *we* have long passed this stage of development, and it is high time for us to be looking forward to a state of things that can endure permanently.

HOME IMPROVEMENTS.

There is but little incentive to the improvement of our homes, so long as there always lurks in the back-ground the probability of abandoning them before long, or selling them at a mere fraction of their value. So long as this feeling prevails, only what is most absolutely needful for the moment, or the near future, and will bring ready cash in the event of a sale, will be looked to by the settler. Thus is destroyed all home feeling—all tendency to the establishment of a home; that home which is so powerful in its influence for good or evil on the young.

So long as our children learn from us to regard our homes merely as the Indian or Arab does his temporary camping-ground—as a thing to be abandoned so soon as we have succeeded in stripping it of its first flush of fertility, by a rapid process of exhaustion by injudicious cropping, without even rest or rotation: so long will they fail to develop in any high degree those social qualities which distinguish the peaceful and civilized tiller of the soil from the nomad. There is in the very plan of existence I have referred to, a degree of selfishness and recklessness of consequences—a sort of "devil take the hindmost" principle, which cannot but leave its impress upon the moral and intellectual life of a community. It has been well and truly said of late by a distinguished philosopher, that after all, our progress and civilization has no more important result to show, nor any more faithful exponent, than the improvement of our homes—including kitchen, pantry, dining room, parlor and bed-room; front and back yard, out-houses, orchards and fields. The correctness of this remark may not strike every one forcibly at first, but a little reflection will demonstrate its strict truth. And, gentlemen, refined, cheerful and pleasant homes are everywhere the marks of an intelligent, refined and moral community—they are practically interdependent, even as, exceptional cases apart, we judge of the *faith* by the *works* which should be its outgrowth.

OUR OLD FIELDS.

What do we see all over the State—what do strangers see in passing along our great commercial highways, the Jackson and Mississippi Central Railroads? I will tell you the first impression of a gentleman I lately met at Grand Junction, who came direct from Texas, *via* New Orleans. Said he: "I don't see how you Mississippi people make a living—either your land is miserably poor, or you have abused it awfully. Why, the whole country along that railroad looks like a turkey gobbler that has been pulled through a briar bush by the tail."

Now, I doubt whether on the whole they do things on a more rational plan in Texas; but then, their lands are fresh, their time has not come yet. But, gentlemen, ours *has*—and it is time we were seeing about it.

Let me tell you another remark—it was made by a federal general, whom we of North Mississippi have reason to remember, and to whom I once applied for protection for the University buildings. He was in a bad humor, having found the country rather too bare of supplies for his men. So, hearing that I was the State Geologist, he inveighed against the false reports that were current as to the productiveness of the soil. Said he: "Talk about the fertile soil of Mississippi! Why, d—n it, it's all like *that!*" and he kicked spitefully at a pile of red sand dug from a cistern near by. You see, he and our friend, the Texan, had quite the same impression on their minds. I leave you to say whether, so far as the highways and older settlements of the State are concerned, it is not fully justified.

From Ripley to Jackson Court House—from Holly Springs to Woodville, do we not find many of our towns and county seats in danger of going "down hill" from the encroachment of the red washes; and their surroundings waving with broom sedge instead of bread crops? Have not whole plantations tumbled bodily into gullies, almost beyond the reach of any practically possible reclamation?—the hills denuded of soil and even subsoil, and the fertile valleys overrun with a flood of arid sand, where but a few years ago we had lively streams running between high banks all the year round—through fields waving with corn and cotton.

THE SAND SCOURGE.

In those portions of the State which are comparatively exempt from the sand scourge, the extent of injury inflicted by it on other districts would hardly be believed. In some portions of Marshall and Benton counties, lately visited by me, it has assumed the character of a public calamity. The turning-out of worn hillside land during the war resulted in the formation of innumerable hillside gullies, which, so soon as the quicksand is reached, at the depth of a few feet, widen with fearful rapidity, not only by washing, but by *caving*. The hard, unprotected and untilled surface of the land sheds a much larger proportion of rain water than formerly, and much more rapidly, into these gullies; from which, during heavy rains, a tumbling, gruel-like mass of sand and water emerges upon the valleys.

Before the war, the then comparatively slight danger of inundation by sand was averted by straightening the channels of the streams, thus increasing their power of carrying off the sand poured into them from hillside rills. But now, the quantity of sand has so greatly increased, that not only have the old valley ditches been filled up, but it is extremely doubtful whether any amount of labor likely to be at the command of private individuals, could have kept, or keep them open. The huge volumes of sand carried down from the hills must lodge somewhere, and ruin somebody's land. Each land owner blames his neighbor above or below him, as the case may be; and the law-suits resulting therefrom would be legion, but for the fact that each one being to blame himself, more or less, for precisely the same omissions, all parties would in the end fare about alike.

There is not, probably, an upland county in the State where we may not find abundant examples of the same general character; resulting in the most serious injury to the uplands, and the same, or worse, to the smaller valleys.

What we see in Marshall, Benton and Lafayette, we find repeated with lamentable exactness in Winston, Lauderdale and Jasper, in

Marion and Wilkinson—wherever, in short, sand is found underlying the subsoil at no great depth, which is the case of fully three-fourths of the uplands in the State.

THE PRAIRIES.

As for the rest—look at the prairies of Northeast Mississippi—Monroe, Chickasaw and Lowndes—the garden spot of Mississippi. Is not the soil, which for thirty or forty years has unremittingly yielded magnificent crops, rapidly giving out? Were not lands that twenty years ago, could not have been bought at $20, offered at $5 per acre just before the war, without finding purchasers? I will not argue about their present market value—it is influenced by many outside circumstances; but through them all glare the stubborn fact, that around the older settlements—say Okolona—hundreds and even thousands of acres of the richest looking prairie are thrown out of cultivation, because, notwithstanding their nearness to the market, the crops they yield have ceased to be remunerative, both from their indifferent quantity and quality, and from their liability to frequent failure.

THE "INEXHAUSTIBLE" BOTTOM.

From the growing infertility of the uplands, the large planters especially have for years past sought refuge in the bottoms, particularly in that of the Mississippi. In these, indeed, many of them have thought they had found that fabled soil that never wears out. They pooh-pooh the idea of its ever becoming exhausted, and in some dark corners, even now pitch their cotton-seed into the bayous, to get rid of it. But the laws of nature are the same every where, and cannot long be defied with impunity. It is true that the soil of the Mississippi bottom is very rich and very deep; and that that of the Nile bottom is still as fertile as it was three thousand years ago. But that is because of the annual overflows, which continually bring down a fresh and very rich soil from the table-lands of Abyssinia.

But as it now turns out, the best soil of the Mississippi bottom (the "buckshot") is *not* an alluvial one, formed by the deposits of the present river: but is analagous to the upland prairie soils. Moreover, we are trying our best to shut out the annual overflows, and will doubtless succeed in so doing. The time of exhaustion of the bottoms, then, will surely come, and will be felt by some of the present generation; the more, as one of the causes of the fertility of these lowlands is that they now receive the cream of the soils of the uplands.

"WE MUST CHANGE ALL THAT."

Can this state of things continue? If it does, how long will it be before the fairest portions of our uplands (which were of course picked out by the first settlers) shall become a mere waste of red gullies and broom-sedge? If we do not use their heritage more rationally, well might the Chickasaws and Choctaws question the moral right of the act by which their beautiful, park-like hunting grounds were turned over to another race, on the plea that *they* did not put them to the uses for which the Creator intended them.

Under their system these lands would have lasted forever; under ours, as heretofore practiced, in less than a century more the State would be reduced to the condition of the Roman Campagna.

This will not and cannot be, in these times of rapid progress, inter-communication and hot competition. If we ourselves were to fail to do by our lands as we ought, they would simply pass away from our hands into others more willing to conform to the requirements of the times, as dictated by prudence and common sense. The farmer who raises a mere fraction of a bale of cotton upon an acre of land, cannot long hold his ground alongside of him who, with little more labor and expense, raises an entire bale, and that of a quality which in the market will bring two or three cents more per pound.

It seems almost trite to say that it will not pay to cultivate poor land; and yet I dare say there are few among us who do not do so to a certain extent. In this evidently our works are not up to our faith.

CULTIVATING TOO MUCH LAND.

What is worse, the poorer the land, the more of it we try to cultivate; vainly attempting to increase the effect of our labor by diffusing it over a larger area.

There is not, probably, in all our practice, an error more widely diffused, and more fatal in its consequences, than this one of attempting to cultivate too much land—tilling it badly of course, tiring the teams by dint of just traveling over it, and manuring it not at all. The labor and expense thus scattered over four acres of ground, to raise one bale of cotton, would frequently, if bestowed on the improvement of half the area, almost double the product; and the staple raised on a more generous soil would be of a much better quality than that gathered off the half-starved stalks on the four poor acres.

One cause of this persistence in the cultivation of poor land probably is, that opinions differ greatly as to what *is* "poor" land; and where no farm accounts are kept, it is not always easy to judge when culture ceases to be profitable. Then, too, our farmers are too much in the habit of considering all lands as "poor," that will not produce profitable crops of cotton or corn.

"COTTON OR NOTHING!"

"A cotton raiser has no time to plow under peas." That is the doctrine promulgated, not many months ago, in one of our agricultural journals, by a disconsolate owner of poor sandy land; who found he could make nothing at raising cotton, yet evidently was determined to raise nothing else—"let justice be done, though the heavens fall."

Fellow-citizens, this "cotton or nothing" idea has been the bane of our system of agriculture. It has aggravated beyond measure those evil effects on the soil which seem everywhere to have attended the labor system to which we were committed before the war. The unceasing repetition of cotton cropping on the same soil, without rotation, return or rest, has exhausted in the course of twenty years, lands that with anything like a rational system of culture, and without any pur chase of manure, should have lasted a century.

When I say exhausted, I do not mean to say that *no* crop could be profitably raised on the soil. By no means. Clover, the grasses and other forage, bread and market crops, will still flourish on many a soil, that will refuse to yield even a fifth of a bale of cotton per acre. But these did not enter into our calculations. It had become traditional that the Northwest was to feed us, the Northeast to supply almost everything else—and it was only when these supplies were forcibly cut off, that we began to realize the folly of the system, and found ourselves somewhat like King Midas before his golden viands: with lots of cotton but nothing to eat!

WHAT IS TO BE DONE?

What then must we do to avoid all these evils? We cannot all go to the "bottom" if we would—we do not want to move West. We have but little capital to invest. How shall we invest what we have?

Well, there is no royal road to the solution of the question, and quack medicines will not help us here.

Keep up your land! That is really *the* great problem, before whose importance all questions of detail sink into comparative insignificance. A generous soil will produce almost anywhere, any crop adapted to the climate. It will repay your labor bountifully even when not bestowed to the best advantage, because of the wide margin it leaves for profit and loss. It will carry your crops safely over a spell of drouth that will singe out of existence those on "poor" land; and if full justice is done to it by deep tillage, it will equally secure them against such excess of wet weather as would be fatal to poor and stunted seedlings, grown on "worn" soil. And last, but not least, this comparative certainty with which the farmer can count upon remunerative crops, enables him to forecast his resources to vastly greater advantage, and with little risk of serious mistakes. He knows what purchases he is justified in making at the time when they will be most advantageous to him; securing him those advantages so highly prized in all other pursuits, of steady and certain profits, with little or no risk; save from extraordinary events not susceptible of being included in forecasts, by either the farmer or any other business man.

FAT YEARS AND LEAN YEARS.

From providential causes, there will always be "fat years" and "lean years," comparatively speaking. But, extraordinary cases excepted, it is our own fault if we have what may properly be called years of famine, as contra-distinguished from those of average harvests. The terrific famines which even in our times ravage oriental countries, result mainly from the fact that in consequence of the very rude and imperfect mode of cultivation practised there, the issue of the harvest depends almost wholly upon the seasons. With improved methods of agriculture, no such total failures *can* happen to entire States in the Occident; any more than to China or Japan, where the principle of the maintenance of fertility alone, even with the rudest implements of tillage, prevents such disastrous events from assuming a general character.

How much more ought *we* to do, with the additional advantages given us by the gigantic progress of the arts and sciences, advancing hand in hand so fast that but few of us are able to fully keep up with their progress! And if our advantages are so much greater, how much greater are also our responsibilities for their proper use!

Yet if we, on this side of the Atlantic, and especially south of the Ohio, have thus far remained exempt from the scourge of famine, it is more owing to the high native fertility of our yet unexhausted soils, than to our own foresight. Improved implements, too, have been so temptingly set before us, that unwittingly and sometimes almost unwillingly, we have in practice adopted some of the principles of Progressive Agriculture, in our Brinly Plows, our cultivators, our cotton gins and presses, our improved varieties of crops and stock. But all the while we have been slow to accept, and slower still to carry into practice, those fundamental principles upon which depends our continued prosperity as an agricultural people, and to the faithful practice of which all other improvements are but auxiliaries.

Taking it for granted on all hands, that the maintenance of the fertility of our fields is the fundamental condition of all agricultural improvement: let us briefly consider what are the principles upon which we must work to accomplish that end, and the means at our command.

THE CAUSES OF EXHAUSTION.

It is now known to all of you, that the exhaustion of soils is owing to the withdrawal from them of the ash ingredients of the crops. If, instead of removing the crops from the fields, we were to turn them under with the plow every year, the productiveness of the soil would continually increase by cultivation; because the soil ingredients taken up by the crops would be returned to the soil in a more available condition, while a partially fresh supply would, besides, be taken in by every successive crop.

If, however, we *fail* to return a portion of the crop, the soil ingredients of that portion must be returned to the land in some other form (*i. e.* that of manures), if fertility is to be maintained.

Practically, there are crops which, so long as only the portion actually sold is removed from the land, while the rest is at once returned, will not sensibly diminish the productiveness of a strong soil, for a great length of time. Pre-eminent among these stands our great staple cotton. Here are the data:

SOIL INGREDIENTS REQUIRED FOR A BALE OF COTTON.

		Pounds Ash.
1,350 pounds of Seed Cotton,	400 lbs. Lint contain................	4
	950 lbs. Seed contain.............	41
		45

Of the 41 pounds contained in the Seed alone:

475	pounds of Hulls contain....................................	9½
368	pounds of Cake contain..................	31
107	pounds of Oil contain about..................................	½
950		41

A corresponding crop of

35 bushels of Corn contains........	25
15 bushels of Wheat contains..	18

This shows what *might* have been done with our cotton lands, had a rational system of culture prevailed from the beginning. It shows what we may still do with our newly opened lands, and with our old ones after they are reclaimed from their present dormant condition.

THE COTTONSEED-OIL MANUFACTURE.

And what are we now doing? Instead of returning the cotton-seed to the soil, we now, to a large extent, sell it to the oil manufacturer—getting in return for *ten times* the amount of soil ingredients removed by the bale of lint, about *one-tenth* of what the lint sold for; or about one hundred times less, weight for weight. Now when it is considered that the soil ingredients so sold could actually, within ten years, have been converted into ten crops of cotton: the operation, in a business point of view, appears simply preposterous.

The soil ingredients are our capital, and it is this we sell, at a mere nominal value, in selling our cotton-seed.

Should we, therefore, discourage the cottonseed-oil manufacture?

Yes, if we continue (as is the case now) to export the cake, and with it the essence of the soil's fertility, to New and Old England. *No*, if we get back the cottonseed cake, leaving the oil as toll to the manufacturer. Seed cake is worth more to us than the raw seed to which it corresponds. But in order to carry out this programme effectually, it would be necessary to have oil presses scattered broadcast through the country. I believe the time will come when our cotton-seed will be carried to the press as naturally as we now carry our grist to mill. But for the present it will be most advisable for farmers to return their cotton-seed religiously to the soil on which it grew.

And this is the first and foremost rule to prevent exhaustion: *Return faithfully to the soil* (whether directly, or through cattle fed *with them,*) *all those portions of crops whose selling price would not enable you to buy back the manurial ingredients it contains*, besides yielding you a full and fair remuneration for the labor of production. And least of all let any portion of such crop go to waste, or be burnt.

MANURE-MAKING.

The utilization of the offal (so to speak) of our crops is most intimately connected with that of the preservation and collection of manure. It seems incredible to those who have never tried it, what a large amount of fertilizing material can be accumulated in the course of a year, when we stop the thousand-and-one leaks through which the greater portion of these ingredients now escapes; when in short *manure-making* is regarded as a business, second in importance only to *crop-making*, of which it is one of the chief conditions of success.

In this we must follow the example of all the older communities in the world, and the sooner the better. Unfortunately, many of those who attempt to enter upon this measure of reform, too frequently stop half way—allowing, *e. g.*, their cattle to run at large, after having fed them on the produce of the farm; and thus scatter abroad on a miserable woods pasture, the manure that should have been faithfully returned to the cultivated soil. We cling tenaciously to the habit of letting our cattle range in the woods, when what they actually get in the way of pasture is not worth half the annual cost of keeping up fences, in the greater portion of the State. The abatement of the fence nuisance is one of the most pressing measures of reform in the more thickly settled portions of the State; yet while many a farmer pays his State tax several times over in fence repairs, he will protest loudly against the infringement upon his rights, that would exclude his cattle from trespassing upon the land of others, or upon the public domain.

Few are aware how small an area, well cultivated, will green-soil stock; saving their manure, and improving their condition in every way. Again, how can improved breeds of cattle be maintained, if allowed to range all the year with all manner of scrub-stock?

OUR MARLS AND GREENSANDS.

Next in importance to the preservation and use of manure, is the utilization of the natural manures found in the State. Of these we have a great abundance, and they will greatly help us to redeem our lands from the consequences of maltreatment it has undergone heretofore.

(Here the speaker exhibited a map showing the marl and greensand regions of the State.)

Here, in the north-eastern portion of the State, is the Pontotoc Ridge, extending from the Tennessee line to Houston, with an average width of twelve miles. The body of its hills consists largely of greensand marls, at many points fully equal to those of New Jersey; alternating with strata of limestone which will, on burning, yield an article of lime worth twice as much for agricultural purposes, as any of the imported article.

Alongside is the prairie region proper, which the Rotten Limestone underlies everywhere. While not equal to the rock of the Pontotoc Ridge for agricultural use, it will nevertheless, when burnt, be still more valuable as a fertilizer than imported lime; and where it will crumble readily, produces a fine effect even in the raw condition, as a marl. It offers a most ready means of rendering available for profitable cultivation, the intractable soils of the adjoining Flat-Woods, whenever railroads shall make cheap transportation available.

In Central Mississippi, we have a marl belt traversing the State, from Vicksburg to Winchester, with an average width of about 36 miles northward of that line. The quality of these marls is, on the whole, superior to that of the Pontotoc Ridge marls; as they contain less inert matter, and almost always a considerable amount of green-sand. Within a few miles of where we stand, there are beds of marl whose manurial value is sufficiently great to pay for shipment by rail to a considerable distance; and they might readily be concentrated (as are the New Jersey greensands) to increase their value, weight for weight.

These marls, in most cases, are not mere stimulants, but true manures; worth more per ton, in some localities, than several of the manufactured fertilizers in the market.

Northward of the marl region, and southward of a line drawn from near Carrollton to Marion Station, we find numerous beds of greensand, of high manurial value as sources of potash; of which substance the deposits near Vaiden and Winona contain from one-and-a-half to two per cent. These materials could readily be concentrated, so as to bear shipment by rail all over the State.

The greater part of these facts are stated in detail in my report of 1860; some of them have been developed by researches made since then. As yet, they have attracted but little attention—in part, no doubt, from a perverse disposition in human nature to depreciate home resources; in part, also, because the manure question is but *now* assuming that threatening aspect which, before long, must rouse up the most indifferent to a sense of the absolute necessity of adopting a more rational practice.

These marl beds can do for us all, and a great deal more than that which marls have done for Virginia and parts of the Carolinas; for the reason that they are more complete fertilizers, and more widely distributed. *Good* marls and green-sands do not, it is true, exist in every locality within the areas above described; but with improved means of communication and preparation, they can be made very generally available, even beyond the limits of their region of occurrence.

I am glad to see that of late, especially in the neighborhood of Jackson, experiments in the use of marls are becoming more frequent.

COMMERCIAL FERTILIZERS.

As for commercial fertilizers, their most proper use applies to two or three cases. First, to make up the deficiency in the annual returns to the soil, arising from the sale of our products. This is in the nature of a common credit and debit account, which should be kept by every farmer; for on the regular balancing of this account depends the maintenance of profitable productiveness.

Secondly, they are profitably used in making up for the *one-sided* wear, so to speak, of our lands; whose wearing-out is frequently dependant merely upon the drain of phosphates; in which case the application of commercial super-phosphates will be followed by a surprising effect. They are therefore highly valuable adjuncts in the reclamation of worn lands, enabling us to make crops while the soil is being rehabitated by subsoiling, fallowing, green-cropping and marling.

But in order to perform these services, they must be what they claim to be; which, as we all know to our cost, is too frequently not the case. I am satisfied that we ought to adopt the same remedy as other States, allowing no fertilizers to be sold, that have not been sampled, and their per centage of active ingredients ascertained and certified to, by a State Inspector.

As a rule, we pay by far too much for them. And in any case, they will not do to rely on, as they supply but one, or a few of the ingredients required by plants. The main return should, as a rule, come from the farm itself.

ROTATION OF CROPS.

Among the most essential points in the treatment of all lands, both manured and unmanured, is *rotation*—an annual change of crops in proper succession.

Without this, the land ceases to produce much sooner, if left unmanured; and manure used is less profitable.

We have sinned greatly against this important principle in our incessant cropping with cotton; and if there were no other reasons why we should diversify crops more than has been done, this one would be sufficient to condemn our system. I remark that the proper succession for some of our important crops is far from being well settled; and that the question, if settled for one class of soils, frequently arises anew with reference to other classes, and will require numerous and well devised experiments for its settlement.

Apart from manuring proper, and rotation, there are a number of expedients for improving and stimulating the production of soils; such as *subsoiling, thorough-drainage,* the *fallow,* and *turning-in of green crops.* These are of especial importance to us in the reclamation of our worn lands, since they enable us to dispense, for the time being, with much of the heavy cash outlay otherwise required for the purchase of manures.

SUBSOILING.

As for subsoiling, its great effectiveness with us depends largely upon the lamentably shallow tillage that has mostly been bestowed upon our lands. We have cultivated large areas badly, scratching, on an average, three to four inches deep; and it is mainly this shallow surface layer that has been worn out, while the greater part of the properly arable soil lies untouched beneath. True, it is not in a condition to bear heavy crops at once; but by the aid of deep and frequent tillage, the fallow, aad turning-in of green crops, as well as, wherever accessible, our excellent marls: most of these worn soils, which were originally "strong," and have not been devastated by washing, can readily be brought back to very nearly their original productiveness. They would, of course, have lasted much longer, had deeper tillage been practised from the outset. For, after all, what we call subsoiling is only deep tillage, without, however, necessarily turning up the deeper portions. *When, where,* and *to what extent* the actual turning up may be practised, depends materially upon circumstances; such as the original and actual depth of the arable soil, its quality, the nature and quality of the subsoil, the conditions of drainage, the crops intended to be raised, etc. It often involves questions of no slight difficulty and complication, which can only be answered either by analysis, or else by long and costly experimenting. Indiscriminate subsoiling has in numerous cases in this State proved the very reverse of beneficial; although much more often, it has resulted in very great improvement.

THOROUGH-DRAINAGE.

Thorough-drainage, or under-draining, is very similar in its effects to to subsoiling, but it goes farther in its benefits. It renders the land easier to cultivate; allows the ground to be worked at all times, within a few hours after a rain; secures to crops the advantages of an early start, of a deep soil well aerated, and therefore protects them against either excessive wet or drouth. Hence it renders a fair crop almost a a certainty, every year, even in very extreme seasons, when all else is a total failure.

Thorough-drainage is a costly improvement, and with us will, for some time to come, be confined to the market garden and truck farm; many of which would do far better to invest more in drain tiles and less in commercial fertilizers. Yet there are thousands of cases in which a few or even a single tile drain, well laid, would in a few years pay for itself, by the prevention of washing, the rapid removal of water from large areas now remaining water-soaked for many days ofter rain, and the facilities afforded in cultivation by doing away with open drains,—so troublesome in cutting up fields, and after all, so poorly fulfilling their intended purpose. Brush and log under-drains, if they are to be effectual, require such frequent renewal, that in the end they prove more expensive than tiles.

THE FALLOW.

As for *fallowing*, or as it is usually termed with us, "*resting*" the land, it is so effectual in restoring soils, not originally poor, to productiveness after severe cropping, that it has, time and again, been claimed as the panacea for the maintenance of fertility. But to be effectual, it must not be practised as is frequently done now, viz: simply throwing the land out of cultivation; for it means *tillage without cropping*, and very thorough tillage, too.

"GREEN-CROPPING."

It is most advantageously connected with the *turning-in of green crops*, raised on it for the purpose of soil-improvement. And with our long growing season of eight or nine months, this excellent improvement might be very extensively carried out, without losing a year's crop; by simply removing crops from the field so soon as they can be harvested, and immediately occupying the soil with another crop, even though it be one of weeds—to be turned under either late in autumn, or (as in the case of rye or oats) early enough in the next season to allow time for a feed or forage crop afterwards.

Of all the crops adapted to this climate, none is so effectual in restoring worn lands as Red Clover, whose deep roots draw up nourishment from the subsoil. Hence, in turning it under, we give the surface soil a fresh supply of nourishment available to shallower-rooted crops, which would not, themselves, have gone so deep.

WHERE, WHEN, AND HOW?

Such are the means by the aid of which we may not only increase greatly the fertility of all our lands, but restore to profitable productiveness most of those "worn" soils whose original thriftiness has been destroyed by cropping without return. But like other remedies and appliances, they must not be used indiscriminately; for then disappointment is sure to follow in many cases.

Manures must be of the right kind, selected not only with reference to the requirements of the soil, but also with due regard to the facilities for collecting, using and procuring, by purchase or otherwise, the different materials applicable to the purpose.

Rotation must be properly adapted, both as to the kind and number of succession, not only to the soil, but also to the requirements of the market.

Subsoiling must be practised judiciously, with a due regard to the nature and condition of the soil and subsoil. Otherwise, so far from being a benefit, it may, for the time being at least, become a serious injury.

Under-draining is rarely amiss, yet there are extensive tracts of land in the State on which it would be at least an unnecessary expense at the present time.

There are soils on which the *fallow* can have little or no effect, and would be, practically, a waste of time and labor.

The *turning-in of green crops* can rarely come amiss; yet there are cases in which it will not pay, unless the crop selected for the purpose is of the proper kind.

To determine all these things properly, judiciously, and to the best advantage, in the immense variety of cases which present themselves, requires not only natural intellect, but an *educated judgment.* I cast no slur upon our farmers when I say, that were it not for the wide margin left them for profit and loss on our virgin soils, the egregious mistakes constantly committed by them even in their attempts at improvement, would often prove very disastrous; as is the case in older communities.

The knowledge required cannot be the gift of nature; it must be attained, both by the study of past experience, which has been condensed into principles, and, very often, by experiments adapted to each particular case. The latter is the method very commonly resorted to by those who distrust "book-farming." Let us consider how far, by itself alone, it is likely to lead to beneficial results.

EXPERIMENTING, AND "STUBBORN FACTS."

The readers of our agricultural journals cannot fail to have noticed how very contradictory are the reported results of experiments made by individuals. No matter how well settled may appear to be some question raised, by an experiment devised for the purpose: we will shortly find some one else coming forward who has likewise tried the thing, and reached precisely the opposite result.

Why is this?—Simply because experiments are questions asked of nature, that must be properly put, in order to receive a rational answer. When this is not done, the old adage, that a fool is answered according to his folly, applies here as elsewhere.

In experimenting we attempt to bring about some natural result under known circumstances and conditions. Now, if we fail to take into account any *one* of the influencing facts or conditions, our questions are foolishly asked, and will be answered accordingly. As it happens, the conditions controlling the results of agricultural experiments are oftentimes so exceedingly complex, that even in the most experienced hands, and with the best resources of science, correct and decisive results can be attained only by long series of trials, executed with the most scrupulous care.

No wonder, then, that such great diversity of opinion prevails among farmers, even upon the most vitally important questions. Let me adduce one as an illustration:

"Will commercial fertilizers pay?"

In a general way, some will; and some will not, for the simple reason that they are utterly worthless intrinsically. The farmer who tries them must with us, as a general thing rely on the manufacturer's certificate and honesty; and some of us know, to our cost, how frail a reed this is in many cases. However much opposed to anything resembling sumptuary laws, we shall in this matter, I think, have to follow the example of other States; in the establishment of an obligatory inspection, by State authority, of this very tricky class of merchandize.

But let us suppose that we have the right thing—say a superphosphate. One man applies it to his worn-out pasture, and finds a wonderful effect—say 300 per cent. clear profit on his investment. He becomes excited, and proclaims far and wide: "Mr. N's superphosphate is the thing for 300 per cent. profit!"

His neighbor tries it for corn or cotton on *his* soil, and barely gets back his money. Naturally, he also gets excited—cries humbug—imposition—collusion, etc. Then there ensues a fight—on paper at least. Each one pitches a "stubborn fact" at the other; and each one is right so far as the *facts* go, while both are equally wrong as to the *conclusions* they respectively drew therefrom.

Let me cite an example from my own experience: At Byram station, below this, we have in the banks of Pearl river two thick strata of excellent marl, separated by a 10 inch band of gray clay. Finding that the bank had been dug into, I was told, on inquiry, that "a foolish fellow" had thought the "stuff" good for manure, and hauled off several loads of it. Unfortunately, he had chosen the gray clay, to get at which he had shoveled six feet of excellent marl into the river!

There was an experiment—and I dare say that whosoever perpetrated it, thinks "marl" a great humbug!

A WORD ABOUT "THEORIES."

And here let me say—farmers talk a good deal against "theories," as connected with "book farming." Yet such experimenters—and their number is large among them—are the most arrant theorizers, if they only knew it. From their extremely limited experience, they jump to the boldest conclusions, and stand up for them like a stone wall; the more obstinately the less they know about the matter.

Take, for example, the old popular superstition regarding the transformation of wheat into cheat, and *vice versa;* which contradicts alike the experience of the farmer with all other plants, and the concurrent results of all accurate observation. How, then, has the idea arisen? Simply thus: When (as is but too frequently the case) the seed of wheat and cheat are sown together, both will come up; but ordinarily the wheat will grow much the faster, and in a great measure "choke off" the cheat before blooming time. But if in winter or early spring the field is subjected to close grazing or cutting, the wheat suffers much more than the cheat, which will bear repeated cutting without injury. In consequence, the cheat gets an advantage, the wheat is set back and mostly "choked off" in *its* turn before it can recover; and thus a crop chiefly of cheat is obtained. Nevertheless, if the same mixed seed be sown the next season, and the cutting or grazing omitted: the wheat may prevail in the fight for existence. Then, old farmers say that the cheat has reverted back to wheat!

Again: In the good old times when all our corn and bacon came from the Northwest, it was a settled opinion that forage crops, grasses, clover, etc., would not succeed in our climate. It had been tried, and they could not get a stand, and the summers killed it all.

My friend Philips then, as now, proclaimed, but in vain, that this was an error; that with proper treatment, not only could a stand be obtained, but the summer heat would be successfully resisted. Nobody heeded him—he was a book-farmer, a theorizer, and so forth; "stubborn facts" proved him to be wrong. *Now* you all know that he was right; and that all the stubborn facts amounted to, was—mismanagement. Yet they had settled the point in popular belief.

We must change all that. If the labor, pains, and expense bestowed by scattered individuals on limited experiments leading to unsafe and contradictory conclusions, were combined so as to be judiciously directed to well defined objects, a vast amount of important knowledge would be elicited.

Experimenting is a science and art—a most difficult one to carry out correctly in all cases. It is no disparagement to our farmers to say, that in a great number of cases, they are unable to say exactly where the shoe pinches, and what experiments should be made; above all, unable to judge of all the precautions required to carry them to a successful issue.

Much of the reproach that has been cast upon science as disagreeing with the results of practice, is owing simply to this want of proper qualifications on the part of experimenters, supposed to be, and esteeming themselves "highly scientific men," yet lacking the comprehensive and well digested knowledge, which alone insures correct results, and above all, correct conclusions.

THE GEOLOGICAL AND AGRICULTURAL SURVEY.

A most important step toward this object was the Geological and Agricultural Survey of the State. Its objects are, to acquire, through competent observers, a knowledge of all the natural facts concerning the agricultural and industrial resources of the State; and also to gather for the general benefit all the experience already acquired by farmers concerning the peculiarities and character of soils. These carefully collated and compared with the results of scientific investigation of the soils, will in many cases at once lead to the most important practical conclusions.

Here is a most important function to be fulfilled by our agricultural societies, lodges of Patrons of Husbandry, etc. Experiments are rarely lucrative: individuals cannot afford them. Societies *can*, for the common benefit.

EXPERIMENTAL FARMS.

Fellow-citizens—I would have, with this view, not one experimental farm at Oxford or Jackson, but at least half a dozen of them scattered over the State, in the chief agricultural divisions; as many as possible—one to each agricultural society!

The kind of experiments should be based upon the experience acquired by farmers in the region; discussed with reference to the scientific examination of the region, its soils and other resources, by the State Survey; and the mode of conducting them, the special precautions to be observed, etc., should be carefully considered beforehand with the aid of all the light that science and practice can throw upon the subject.

CONDITION OF THE STATE SURVEY.

I am often asked: Is the Agricultural Survey completed; if not, when will it be?

That depends altogether upon what it is expected to do for the development of the natural resources of the State. I can *now* tell you, or any stranger wanting to settle, what minerals or other useful materials *can*, and *cannot*, be found in the State, and *where*. I can give him a *general* idea of every county, or part of a county, he may ask for, as to kind and quality of soil, timber, water, etc., so far as traveling the usual routes, and visiting all the chief settlements and other points of interest indicated by inquiries along the routes, will give such knowledge.

The gaps existing when my report of 1860 was published, have, in a great measure, been filled up by the gentlemen who have since carried on the work; and a report supplementary to, or comprehending the former one, should be published as soon as all the office work is completed.

(Here the speaker exhibited and explained a map of the State, colored so as to show the geological and agricultural divisions.)

But while while I am able to give all this general information, I am still in a great degree unable to answer vitally important questions of detail. Without the information given by the work of the survey, we should be altogether in the dark, even as to the questions to be settled and the experiments required for that purpose. As it is, we are able to ask questions intelligently, and with reference to the actually existing state of things in each section of the State. But the practical answers to these questions, while foreshadowed by the scientific work done, still require the test of practice, and this test I propose to make by the co-operation of the Agricultural Societies with the State College of Agriculture, where instruction is based upon the extensive collection representing the geological, agricultural and industrial features of the State—as they ARE, and not as they might be supposed to be, from an outside point of view!

For years this has been my pet scheme. A chair of Instruction in Agriculture was the starting point of the Agricultural Survey; for it is obvious that a knowledge of the actual features of the State was absolutely essential to truly practical instruction of youth. In the absence of a pressing call for such instruction on the part of the people, at that time, the original character of the department was gradually lost sight of; and until recently, when the Congressional donation for the purpose of industrial education has revived the interest in this matter all over the United States, though residing at the University, I have had no opportunity of communicating, as a teacher, the results of the State Survey. The publication of my report of 1860, it is true, made the facts accessible; but I mean no disparagement to the older generation, when I avow my conviction, that it is mainly through the *young* men, and through the medium of direct verbal instruction, and not through printed reports carefully put away on their fathers' shelves, that the results of the Survey, and the logical consequences flowing therefrom as regards agricultural practice, will ever become incorporated into popular consciousness.

THE AGRICULTURAL SOCIETIES.

Second in importance in this respect only to the education of youth, is the social and intellectual influence exerted by the Agricultural Societies.

It is there that not only the fathers are brought in direct contact with the progress of the science and art of Agriculture; but there also the sons find the opportunity for applying, for their own benefit as well as that of others, the principles and facts they may have received at the Agricultural College; and for continuing their own studies.

Here are the centers from which correct agricultural practice will spread most readily, the precepts being enforced by example; and the example given by men against whom the mad-dog-cry of "book-farming" cannot be raised.

It is strange and sad, but nevertheless true, that in too many cases, a kind of jealousy has arisen between the Agricultural Colleges and Societies. Since both have in view the same ultimate object—the improvement of agriculture—the antagonism referred to can but be the outgrowth of misapprehension (on one or both sides, as the case may be) of their respective objects, and proper sphere of action. Neither can replace the other, but each is the complement of the other.

INSTRUCTION, AND EXPERIMENTING.

Two entirely distinct functions have generally been attributed to Agricultural Colleges in this country, namely: Instruction in the principles and practice of Agriculture; and simultaneously the performance of agricultural experiments. There is not really any necessary connection between the two functions; and the question arises whether they can without detriment be conjoined. It is my opinion that in not a few instances, the educational interests have suffered by being subordinated, or even too strictly co-ordinated, to the experimental work; and that to this circumstance is owing some of the dissatisfaction with which the results of that education have occasionally met at the hands of the agricultural public.

Experimenting is, in general, the reverse of lucrative, unless conducted on a very limited scale, and with great judgment. An *experimental* farm cannot properly serve as a *model* to be followed, either as to the kind of operations, or their lucrativeness.

It is not, therefore, the best example or field of action for learners, who are of necessity unable to make the proper allowances, and are too apt to carry with them into their home practice, an unwholesome predilection for experimenting. The experimental portion of a College farm can therefore serve for general instruction to a limited extent only, and is rather an expensive institution; nor, in very many cases, can experiments be properly carried out, if hampered by the requirements of instruction.

EXPERIMENTAL STATIONS.

These difficulties have been so well recognized, for some time past, in Germany, and the importance of performing the same experiments in localities differing in soil, climate, etc., is so obvious: that experimental stations, devoted exclusively to experiments of general and local interest, and quite disconnected from any educational functions, have now become numerous all over that country. Their operations are under the same general control as the Agricultural Colleges themselves; and the latter frequently furnish from their *advanced* classes or graduates, competent men as directors and assistants.

Some of the men most distinguished in agricultural science, occupy positions at the head of such establishments.

So well has the distinction between the educational and experimental functions attributed by the act of Congress to our colleges, been recognized, even in this country, that at the first general convention of

Agricultural Colleges, at Chicago, in 1871, a special committee was appointed to memorialize Congress and the State Legislatures on the importance of establishing numerous separate experimental stations in the United States.

In this very direction, a wide field of usefulness is open to our agricultural associations. Even with means much more ample than are likely to be at the command of our State Agricultural College, it can perform the duties of an experimental station to a limited extent only, *because it is in but one place.* Its proper province is mainly, in my view, to determine questions of a *general* character, and also such local ones as its location enables it to act on intelligently and successfully. But wherever located, it cannot possibly test *experimentally*, questions dependent upon local peculiarities of soil and climate other than its own; nor even, without great expense, such questions as those concerning the practical value of our natural fertilizers, marls, greensands, limestones, etc.; whose transportation in adequate quantities would be cumbrous. And then their efficacy would not, after all, be validly tested, since the experience gained might, in many cases, not be at all applicable to the soils of the districts chiefly interested.

CO-OPERATION BETWEEN THE AGRICULTURAL COLLEGES AND SOCIETIES.

Now, it would be perfectly easy for local agricultural associations to carry out such experiments as these, for which the data, suggestions as to the mode of conducting, and analyses or other scientific work required in their course, would be furnished by the central establishment at Oxford; which is in possession of the specimens and field notes giving minute details as to every portion of the State visited. And where examinations had not, in the first instance, been sufficiently minute, or new questions had turned up: it would be the business of the same office to have such supplementary examination made, at the suggestion of agricultural societies. On the other hand, the pupils of the agricultural department would carry with them to all portions of the State, and to the societies of which they would become leading members, a detailed knowledge of the results obtained by the Agricultural Survey regarding the resources of their section—suggestions as to important experiments based on such results; and would be foremost in detecting errors or omissions. They would be pre-eminently qualified to conduct or direct experiments, and report the results for discussion to the society as well as to the central office at the State University.

It is in this way, gentlemen, that we propose, by a close, constant and cordial co-operation with the agricultural organizations of the State, to diffuse what knowledge we already possess, both of principles and facts; to increase that knowledge by experiments and observations furnished by all those interested in progressive agriculture; and in the end to obtain a complete knowledge of the State, its capabilities, and resources, at the command of a community of educated agriculturists, able and willing to use to the best advantage, according to the lights of science (which is but accumulated experience), the gifts bestowed by nature upon our favored State.

THE COURSE OF STUDY.

There is a prejudice, prevalent to some extent, attributing to Agricultural Colleges a vicious tendency to turn away the minds and tastes of students from agricultural pursuits. How far this objection may apply in our case, you will judge for yourselves, by a perusal of our proposed course of study.

(Here the speaker exhibited a synopsis of the Agricultural course, as prescribed at the University of Mississippi.)

This, observe, is the *least* we think an educated farmer ought to study, by way of preparation for his profession. He can study as much *more* as he pleases, or his time may permit. For instance, he should be thoroughly master of his own language, and hence the study of *English branches* goes through the entire course. But we do not *impose* upon him any other language, ancient or modern, since, although useful, they are not essential to him in his pursuit.

Mathematics is pursued to the extent necessary to enable the farmer to do his own surveying, leveling, etc.; to insure correct practice in laying off, and building; and to understand the principles of Physics and Chemistry.

A knowledge of *Botany* should be possessed by every one whose business of life it is to make plants grow. It is among the first subjects taught, embracing, of course, a special course on plants important to Agriculture.

Zoology, embracing also a special course on the animals useful and injurious to Agriculture, succeeds botany in the second year. The propriety of this study is self-evident.

A course of general *Physics*, during the first term of the second year, prepares the student for that of *General Chemistry*. The latter dove-tails with a special course of *Agricultural* and *Economic Chemistry*, during the third or junior year. The latter course embraces, besides the application of chemistry to agriculture, the principles of the arts directly relating to domestic operations; and would be equally useful to students of the softer sex, whose influence on their male competitors is everywhere acknowledged to be extremely desirable. We shall always be glad to see them.

Mineralogy and *Geology*—the sciences treating of the immediáte ingredients and mode of formation of soils, can nowise be omitted; they are taught during the third year. So also is *Meteorology*, or "What we Know about the Weather."

The fourth year is, of course, devoted to the higher branches and finishing touches of the several subjects. Of non-agricultural subjects, we have here, besides the *English Literature* course, that of *Ethics* or Moral Science, which is as necessary to the farmer as to any other profession; and that of *Political Economy* and *Governmental Science*, whose importance is manifest.

Throughout the four years, we have daily lectures in *General* and *Special Agriculture;* embracing, successively, a General Compend (based on Allen's Farm Book); then the details of *Tillage*, *Subsoiling*, *Drainage*, *Preparation of Land*, *Seeding*, *Cultivation*, *Harvesting*, *Storing of Crops*, in general. *Details of the Culture of the Several Crops*; *Horticulture*, *Truck Farming*, *Stock and Dairy Farming*. Finally, during the last or senior year, *Rural Engineering and Architecture*, *Landscape Gardening*, *Rural Economy*, the *General Policy of Culture*, and a Summary of the Course. Last, but not least, the *Special Geology* and *Agriculture of the State*, illustrated by the extensive collections of the Geological and Agricultural Survey, will show in detail, the application of the facts and principles studied before, in each particular portion of the State.

When you consider that, besides all this, the student will almost daily witness the actual performance of farm operations, and take part in them to the extent presently to be explained: I do not see how a tendency to wean him from agricultural pursuits, can possibly be attributed to our course. We shall put the pursuit of agriculture before him in all its manifold bearings and connections, as a profession of a high order, involving no little intellectual capacity and culture, and dignified far above many of those that are now so often resorted to by our youth.

PRACTICAL EXERCISES, AND FARM LABOR.

Now as regards the practical exercises in agricultural operations, there has been a great diversity of opinion as to the extent to which they should occupy the students' time. In the early times of Agricultural Colleges, it was thought that large "model farms," whose labor should be performed by students, were the best means of imparting truly practical knowledge. The first attempts of this kind, probably, were made in Germany; and it was there, too, that the system was first found to be a failure. Since then, the same experience has been gone over in England, France, and in this country.

The result substantially was, first, that the graduates from model farms too frequently failed as farm managers elsewhere; second, that the patronage of these institutions, large at first, rapidly diminished after the first few years, and in some cases was almost reduced to zero.

It is not difficult, at this time, to see the causes of these failures. The performance of so large an amount of physical labor left the students too little time, and still less of the needful mental freshness, to acquire that knowledge of the natural causes governing success or failure in agricultural pursuits, which alone can lead to the correct application of general principles.

They therefore became mere apprentices to the *art* of agriculture—mere routine men, who could do admirably on the spot, and under the same circumstances, under which they had learned their lessons; but were apt to fail so soon as removed from that narrow circle. No wonder then that they themselves, as well as their parents, should come to the conclusion, that after all the home farm was about as good as the model farm; especially considering that the work done at home paid the laborer himself, while that done at the "College," seemingly at least, served mainly to defray the expenses of that establishment.

There is another and often well-grounded cause of failure, and of distrust on the part of agriculturists, of the efficacy of "model farm" teaching. A truly model farm must be lucrative in its financial results; for that, after all, is "what we are after." If not lucrative, it is not, *as a whole,* a model to be followed; and it is clear that to *be* lucrative, all considerations of an educational character must be subordinated to the business requirements. That is, the interests of the *farm,* and not those of the *student,* must be mainly kept in view.

Now, this is not flattering news to those who, in depriving themselves of their sons' labor on the home farm, *often* do so at a considerable sacrifice. Moreover, *should* the model farm be a financial success, it will always labor under the suspicion, that this measure of success is largely owing to its being backed by outside advantages and capital, which would not be at the command of a private farm similarly managed. And the truth is, that this is really the case in most instances, and *must* be so if it is to be of any use as an educational institution.

Under the *pressure* of these objections, and resulting failures, the "model farm" system on the old plan is rapidly giving way everywhere before that system which, while affording abundant opportunity to the student to become an expert in all kinds of agricultural operations, directs his attention chiefly to the *principles* upon which a successful practice must be based, and which are applicable everywhere and always.

We intend that, so far as our resources go, our college farm shall be a model as to the *manner* of doing things. We intend to have the best of its kind, of every implement, crop, or breed of cattle; so that the student may find all that he is taught in the lecture room, exemplified in practice in the best manner, so far as *kind* and *details* are concerned. But we shall make him distinctly understand, that he is not to copy stupidly the whole of what we have and show him; any more than a student of medicine should pour down the throat of his patients the whole *materia medica* of a drug store. We shall educate his judgment as to *when, where* and *how* to apply the improvements we exemplify in order to illustrate the principles we teach; and thus, and thus only, shall we fulfil both the letter and the spirit of the agricultural grant, intended to "promote the liberal and practical education of the industrial classes, in the several pursuits and professions of life."

Neither the letter nor the spirit of that law would be satisfied, were we to return to you your children as mere skilled apprentices to the *art* of agriculture. It is not, and never can be, the object of such institutions to teach the millions their trades; and if they were to attempt such a thing, the millions would not patronize them. They will not pay board for their children for the sake of such instruction as they can obtain by merely imitating the practice of skilled mechanics; because they can do better by apprenticing them to the latter, with little or no expense to themselves.

It is only those who look higher, namely, for an *education* for their children, that will impose upon themselves such sacrifices; and it is for the benefit of such that the donation can mainly serve *directly*. Indirectly, all classes will be benefitted by it; for even those who do not go beyond the imitative process of apprenticeship, will soon fall into the more correct practice of their more successful neighbors.

A few thoroughly educated farmers in each neighborhood or county, will very soon leaven the entire farming community, and to a much more beneficial degree than a more diffusive, but less thorough, instruction to a larger number could attain. And this is especially true where, as I hope will soon be the case all over this State, the agricultural societies form a network of arteries, through which all improvement and useful information can be most rapidly and effectually conveyed to the whole body of agriculturists.

MORE ASSISTANCE NEEDED.

To carry all these things into effect, we, of course, need assistance from the State, such as has been given by all other States accepting the Congressional donation. The latter is expressly so restricted in its use, that the *plant* of the institution must be provided for from other sources. We shall soon have under fence about a hundred acres on the University section, affording a variety of soil, and fine sites for the buildings of the farm, and for the botanical department. But for these buildings and their contents, we need further assistance. And we have no doubt that it will be given.

But above all, we need the co-operation of the *people*—of the farming population; and we believe that great good will come of it, and that it will serve to establish, on a lasting basis, the agricultural and industrial prosperity of the State.

www.ingramcontent.com/pod-product-compliance
Lightning Source LLC
LaVergne TN
LVHW020741120826
845150LV00010B/2291

9781418195069